FACING 2030
COPING WITH **CLIMATE CHANGE**

PETER J. DWYER

Balboa Press books may be ordered through booksellers or by contacting:

Balboa Press
A Division of Hay House
1663 Liberty Drive
Bloomington, IN 47403
www.balboapress.com.au
1 (877) 407-4847

Because of the dynamic nature of the Internet, any web addresses or links contained in this book may have changed since publication and may no longer be valid. The views expressed in this work are solely those of the author and do not necessarily reflect the views of the publisher, and the publisher hereby disclaims any responsibility for them.

Any people depicted in stock imagery provided by Getty Images are models, and such images are being used for illustrative purposes only. Certain stock imagery © Getty Images.

IISBN: 978-1-5043-2097-9 (sc)
ISBN: 978-1-5043-2098-6 (e)

Print information available on the last page.

Balboa Press rev. date: 02/26/2020

BALBOA.PRESS
A DIVISION OF HAY HOUSE

FACING 2030

Coping With Climate Change

Author: Peter J. Dwyer, Research Sociologist, retired Honorary
Fellow, University of Melbourne, Australia

CONTENTS

1. FACING THE FUTURE

Introduction

Currently, there has emerged a widely-held view that the human race is living in troubled times, exemplified in particular by the issue of Climate Change. The following pages attempt a reflection on this to explore how solutions can be developed to the problems we face. Hence the title of this opening chapter - *Facing The Future*. After this *Introduction*, the chapter is divided into sections. The first section is titled *The Passage of Time*, and makes use of historical material concerning life on the planet Earth. The second section, titled *When Times Get Hard*, examines the kind of decision-making process that is needed to resolve our current differences about the problems. The final section of this opening chapter presents a list of the problem-solving elements flowing from the previous sections. Six further chapters then expand on the ideas uncovered in the *Introduction*.

We cannot ignore the fact that towards the end of the year 2019 there was an escalation of the dramatic climatic challenges taking place across the world. For example, in addition to the persistent melting of the Polar Ice-Caps, higher-than-usual temperature differences between the two sides of the Indian Ocean led to dramatic climate contrasts: flooding and landslides in East Africa, which killed dozens of people and forced hundreds of thousands from their homes; while thousands of miles away in Australia, a period of hot, dry weather led to a spate of widespread bushfires across this vast land. By early 2020 at least 19 people were dead, 29 more were missing and more than 1400 homes had been destroyed as fires burned through over five-million hectares of land. Millions of animals were dead and hundreds of thousands more would perish over coming days as a result of killer bushfires terrorising southeast Australia. Native wildlife and agricultural livestock were among the fatalities, with already-endangered species at greater risk of extinction. The extent of the carnage may never be known. Both of these weather events were linked to something meteorologists refer to as the Indian Ocean Dipole. (IOD). The contrasts may be extreme, but they are still not convincing enough to win over the self-proclaimed Masters of the Universe and their fellow climate sceptics, and so the rest of us are left to flee for safety and look for answers by ourselves.

The Passage of Time

From a human perspective how do we make sense of the strange events associated with Climate Change? To do this we need to take it seriously and resist the temptation to consign it to the Too Hard Basket. To achieve this we need to admit to ourselves that what happens as a result of the passage of Time is of immense importance to the future of all of us earthlings, even though in recent times we have become much more pre-occupied with the 'conquest' of Space, both on and off this wonderful planet of ours.

A bit of history can help us here, because, after all, it is an important ingredient of Time! Establishing a timeline is useful for this.

Human scientists seem to agree that what they know as Space and Time began with a Big Bang in the midst of Chaos some 13.7 billions of years ago. It is important to notice that scientists have also recorded dramatic climatic events since that time, with as many as five Ice Ages during the Earth's history. The first occurred 2.4-2.1 billion years ago and before any humans existed, but the most recent, called "The Ice Age", impacted humanity and reached its peak only about 18,000 years ago before giving way to an 'interglacial epoch' 11,700 years ago. So, of course, Ice Ages have often affected the Earth, and thus are part of its 'Time', as the following list shows.

First Ice Age	**2.4 billion years ago**
First Life-forms	2 billion years ago
Cryogenian Ice Age	**720 to 630 million years ago**
First Ape-forms	20 million years ago
Early hominids	1.7 million years ago
Homo Erectus	400,000 years ago
Homo Sapiens	**200,000 years ago**
Last Neanderthals	40,000 years ago
Last Ice Age Peak	**18,000 years ago.**

What this time-line reveals is that climatic disruptions must be accepted as integral to our history. It uncovers for us the most likely explanation for the fact that global warming seems to have caught us all by surprise. It reveals that for us the Earth's climate problems are 'unexpected' because, after the passing of 'The Ice Age', dominant humans felt that they had now outlived any cataclysmic threats to their own future, and therefore could indulge in a preoccupation with 'Space' - crossing new frontiers, conquering new territories, building huge empires and utilising all available resources to facilitate the subsequent spread of population and thus become the dominant species on Earth, as the following list shows.

10,000BC	1 million humans
9000-7000BC	**Neolithic Revolution** (farming settlements)
2000BC	27 million humans
200 AD	200 million humans
1500 AD	400 million humans
1730-1850 AD	**Industrial Revolution (Age of Enlightenment**)
1850 AD	1.17 billion humans
1925 AD	2 billion humans
2000 AD	6 billion humans
2019 AD	**7.7 billion humans**

These massive numbers demonstrate that climate change is not the only major problem affecting our world – there are after all our massive population problems, like the worldwide refugee crisis and ongoing child mortality rates (or mounting problems like urban sprawl or even tourism), but there is so much history behind these figures that they serve as a 'timely warning' not to just push to one side other problems like climate change.

It is possible, for example, that the figures contain a hidden message: that if we are to act, before it is 'too late', we need to think both backwards and forwards at the same time. Thinking of the past (history) should always inform attempts to predict the future (forecasting). This avoids the mistake of dealing with Time in a more linear fashion – as a kind of trajectory, like a Spatial journey from point A to point B, which tends to reduce Progress to a process of 'leaving the past behind'. Our climate and population problems are clearly directly connected with this dominant 'linear mindset'.

There is a further major insight here. It is that a Janus-like experience of Time helps people to be happy to live with and work towards a comprehensive resolution of unresolved dilemmas, whereas a more linear approach leads to resolutions of dilemmas by seeing them as an either/or challenge (see Lewis-Williams, p.58; and Burling, *The Talking Ape*). The third chapter of my 2010 book, *Recovery*, went into some detail about this to demonstrate how when we are confronted with a contrast between two poles of reality or thought, we come up with the answer by opting for one pole and rejecting the ambiguity introduced into our consciousness by what we now define as its 'opposite'. The assumed opposites become dichotomies - left versus right, regulation versus free markets, male versus female, science versus art, West versus East, race against race, creationism versus evolution, and us versus them.

These misleading dichotomies often arise because we ignore a final element residing in our consciousness which serves as a kind of irritant for some and a source of inspiration for others — a sense of wonder or even mystery. The really big mystery for our particular period of history would be to find ways whereby the current human preoccupation with the Conquest of Space could be used as a parallel to revitalise in us a matching concern for the Passage of Time. There is in fact an obvious parallel - one offered in 1970 by an anthropologist confronting the challenges of the late 20th century - Margaret Mead. In a book titled *Culture and Commitment*, she coined the phrase 'Immigrants in Time', and used a Spatial analogy to explain an Earthly passage of Time.

> Mead's insight was based on an analogy between the experience of migrant people moving from the familiar surroundings of their native land, and its culture, into a new and strikingly different land with its seemingly strange customs and standards of behaviour. Theirs was an experience of strangeness in the sense of being 'out of place' in their new surroundings. Mead suggested that, since the 1940s, adults of every nation had been confronted with new customs and standards in their own country as a result of the process of change. Theirs has been an experience of strangeness in the sense of being 'out of phase' in a new world which demanded that they learn new ways of doing things and find new answers to old questions (Dwyer, *Recovery*, pp.129-30).

Using this insight of comparing processes of Space and Time, we could develop for ourselves a genuine sense of learning from the past. This could help us cope with problems (of our own making) which, strangely enough, give rise to our doubts about how to deal with Global Warming.

If we were brave enough to attempt this we would soon recover a sense of being 'custodians' of the planet. Adopting this type of perspective would help us to escape from a trap hidden in the claim made by financial experts in this century that humans are now the Masters of the Universe, thus exaggerating a common misreading of the nature of progress. The reality is markedly different, as Ronald Wright has pointed out:

> Civilization is an experiment, a very recent way of life in the human career, and it has a habit
> of walking into what I am calling progress traps (Wright, *A Short History of Progress*, p. 108).

To avoid any such progress traps, we definitely need to learn from our past. For example, we could learn from how humans reacted to the Last Ice Age some 30,000 years ago. They became cave dwellers, leaving for later generations, all over the planet, a lasting heritage in the cave paintings (like the Ibex at Cougnac in South-West France) that humans now so often frequent.

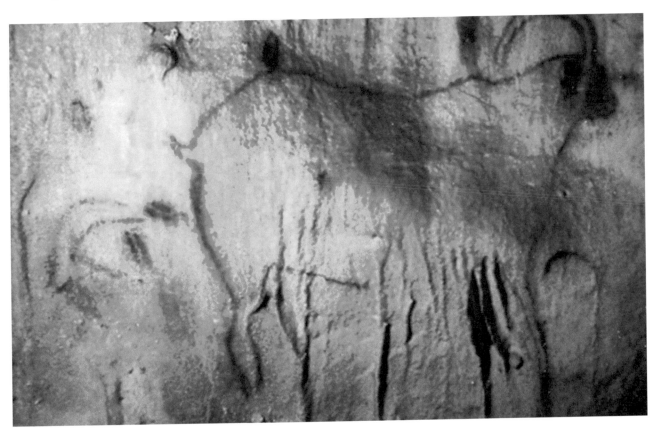

In France, for example, about 30,000 years or so, Paris would have been in a polar desert unfit for humans or animals, whereas the south-west still had both humans and animals in suitable habitats. People got on with their lives and fostered a symbiotic relationship with the animals which they then portrayed in a variety of ways on the walls of the caves they utilised. However, learning from the past is very different from dictating about the past, as the initial theorizing about cave art unfortunately demonstrates.

> Those earlier theories were encapsulated in the summation by Abbe Breuil, a recognized expert, that what the cave art told us about the society was that the chief aims were that 'the game should be plentiful, that it should increase, and that sufficient should be killed'. These misleading assumptions about 'man the hunter' dominated the field…. Later careful examination revealed that 'the majority of the animals depicted in cave art are shown in peaceful attitudes', and many of the animals depicted were not the ones that were hunted, and in fact those which formed the staple diet in particular regions were noticeably absent (Dwyer, *Recovery*, pp. 59-60).

In Australia, there are also similarities in the lessons we can learn.

Thus, in Australia, which was inhabited by its first peoples over 60,000 years ago (before The Ice Age and its peak 18,000 years ago), there are many sites bearing the paintings of local tribes, which have their own style, with 'Mimi' or stick-figures (like some art in Spain) and a kind of 'X-ray art' of the internal organs of the creatures pictured, but it soon becomes clear that a sense of harmony acts as an underlying theme in their story-telling:

> the essential similarity and the essential oneness of all living, natural things, of which human beings are part; all share a common life essence, a kind of basic identification with the natural environment and all it contains (Berndt, Berndt, and Stanton, 1982, p. 21).

Once we make the effort to combine these two elements (becoming 'custodians' again, and respecting our heritage) there is a third step we need to take if we are to restore a sense of Time to our mindset about the world we live in. What we also need to do is to be honest with ourselves and begin to acknowledge the mistaken paths we humans are so liable to follow. Or, to quote Ronald Wright,

> The great advantage we have, our best chance for avoiding the fate of past societies, is that we know about those past societies. We can see how and why they went wrong. Homo sapiens has the information to know itself for what it is: an Ice Age hunter only half-evolved towards intelligence; clever but seldom wise.

> ....We have the tools and the means to share resources, clean up pollution, dispense basic health care and birth control, set economic limits in line with natural ones. If we don't do these things now, while we prosper, we will never be able to do them when times get hard. Our fate will twist out of our hands. (Wright, R. *A Short History of Progress*, pp. 131-2).

It is not simply a matter of admitting the mistakes that our fellow humans might have made in the past, but of also accepting responsibility for the mistakes that might flow from our current mindset. For example, the 2019 fires ravaging the Amazon indicate that, even though we might 'have the tools', we are yet again following a mistaken path, and inevitably making things worse, not better. There is still a lot of learning to do.

When Times Get Hard

There is a final question we must ask to make sense of what we need to do about climate change. Is there some kind of timeline for the human race when, like now, times get hard?

The responses of the younger generation in recent times provides us with some kind of answer, because it suggests that they have learnt, from our failure to act, that everybody (themselves included – even though they are still only children!) has a part to play if the human race is to unite as 'custodians' of their world and care about the living pattern of Time. We humans now have of a kind of 'tunnel vision' view of the future which may be an unfortunate result of a new attitude that humans adopted as part of what they saw as the Age of Enlightenment. The 19th century French intellectual, de Saint-Simon, spelt it out.

> Poetic imagination has put the Golden Age in the cradle of the human race, amid the ignorance and brutishness of primitive times; it is rather the Iron Age, which should be put there. The Golden Age of the human race is not behind us but before us; it lies in the perfection of the social order. Our ancestors never saw it; our children will one day arrive there; it is for us to clear the way (de Saint-Simon, 1814).

This presents us with a likely contrast between an accepted modern human assumption about rejecting the past and looking for a Golden Age, versus the proven value in former times of actually uniting to learn from the past – a sense of a living history. It is worth noting, therefore, that there really are times when humans willingly share a sense of 'common purpose' (see Judt T., p.63). There is much historical evidence to show that when times got hard in the past humans often displayed an inclination to pull together, and in the best of times even relied on the marginalised to join what was seen as the 'common fight'. Unfortunately, this is harder to imagine these days because the prevailing wisdom is that 'the past is past', and that those who 'clear the way', e.g. lobbyists and the powerful, have a head-start on the marginalised (people like Brazilian tribes trying to protect the Amazon forests from being 'cleared' - which is also now part of the Global Warming story).

There is a strange irony in all this, because of the widespread success last century (in both Latin America and the rest of the world) of Paolo Freire, a Brazilian educator! In the light of his success, we have real evidence that the struggles of the marginalised teach us about how to deal with Hard Times. His book, *Pedagogy of the Oppressed*, says it all. His ideas were adopted elsewhere throughout the world because his starting-point was to set up an interactive dialogue with the marginalised for the purpose of restoring active voice to people who felt that they had been excluded from any genuine participation or future in their own societies. He firmly believed that "cooperation can only be achieved through communication", and that the resultant dialogue with the marginalised "does not impose, does not manipulate, does not domesticate, does not "sloganise" (Freire, p. 168).

This approach is shared by the marginalised first peoples of Australia, as Pascoe has said.

> We will approach the idea of One Nation not by exclusion, but by an inclusion that rarely gets mentioned, Aboriginal participation.
>
> More importantly, however, it will have intellectual and moral benefits, freeing us from the mental gymnastics we currently perform to rationalise colonialism and dispossession….It seems improbable that a country can continually hide from the actuality of its history in order to validate the fact that having said sorry, we refuse to say thanks. (Pascoe, B., p. 228).

If people give up their mental gymnastics, and cooperate in facing up to hard times they may finally dispense with the Too Hard Basket which has served them so badly in recent times. Think about it!

Summation

We can conclude with a short list of the problem-solving elements flowing from our discussion. The following five elements are important.

1. conduct regular participatory reviews, including interactive dialogue with the marginalised.
2. think both backwards and forwards at the same time. Thinking of the past (history) should always inform attempts to predict the future (forecasting), but remember that learning from the past is very different from dictating about the past.
3. build on a sense of being 'custodians' of the planet, instead of acting as self-styled Masters of the Universe.
4. acknowledge the 'mistaken paths' humans have taken over time.
5. dispense with the Too Hard Basket.

Tomorrow is another day.

2. THE HUMAN CLOCK

As our human clock moves forwards through the 21st century, it would appear that, in the eyes of a younger generation (such as The Hong Kong protestors in 2019), those who claim responsibility for determining their futures seem in fact in many ways to have gone backwards rather than forwards in time. It is worth asking whether we are turning the clock back even further than we realise. We seem to be reverting to the old antipathies and simplistic dogmatisms of the Middle Ages, as each group or nation appeals to its particular god to justify its own inhumanity. A series of constraining dualisms confronts us, and some abstract predetermined slogan or principle is invoked to justify competing claims to what is true and right. You are meant to take sides. Instead of doing that, I want to draw on some of the ideas I have already grappled with in my recent 2010 book, *Recovery*, and use those ideas alongside some more recent information.

One clear insight about this is revealed by the fact the Oxford Dictionary chose the word "toxic" as it's 2018 'word of the year'. It argued that "toxic" had "truly taken off into the realm of metaphor, as people have reached for the word to describe workplaces, schools, cultures, relationships and stress". It added that the "Me Too" movement had "put the spotlight on toxic masculinity" whereas in politics more broadly "the word has been applied to the rhetoric, policies, agendas and legacies of leaders and governments around the globe."

Given its toxic aspects, our current world-wide crisis could well be read in another more historical sense. It could be read as proof that we have entered another of the great periods of Cultural Revolution in the history of human civilisation, with the prospect of actually learning from our past instead of simply repeating it. We can acknowledge that there are definite similarities with the previous cultural upheavals that our predecessors actually survived, which suggests that there may well be lessons to be drawn from the past about how the human race managed to deal with the problems facing it.

For a start, we can recognize that this type of revolutionary process involves a radical shift in people's view of the world and their expectations of life. It is based on a re-definition of the values people consider paramount and is manifested in a clear re-ordering of the priorities that determine the way they organise their personal lives. At a more systematic level it involves a three-fold process: a critical analysis of the prevailing culture to discover its dehumanizing elements; concrete programs of action to lessen the personal and social costs that the prevailing culture demands; and thus to restore particular human values that have been sacrificed to enable that culture to prosper.

That is why in the past these transformations were even spoken of in revolutionary terms. Two of them in particular stand out. The first is the Neolithic Revolution of pre-historic or pre-biblical times, which featured the shift from nomadic hunter-gather tribal groupings to agricultural societies based on permanent settlements, cities and the formal organisations and institutions that we take for granted as essential to civilised society in our own times. However, the transition was not only a gradual one lasting many thousands of years between about 9000BC to 7000BC and later, but it was not absolute in its scope. Thus, even in our own time, nomadic hunter-gather tribal groupings still exist throughout the world — even in those areas of the world such as the Middle East and Africa where the great 'revolutionary' civilisations of the ancient empires of Egypt and Persia had supposedly consigned them to a forgotten past. Furthermore, the actual process of transformation incorporated many of the pre-established traditions, customs and social practices of nomadic tribes, by re-shaping them to suit the needs and demands of a more settled form of human organisation.

The second major transformation is referred to as the Industrial Revolution. Here too, a picture similar to the Neolithic Revolution emerges. The time sequence was shorter — hundreds not thousands of years — but it was still a drawn-out process of upheaval as villages and cities became the focal points for a transformation from the feudal modes of social organisation derived from an agrarian way of life into the industrial society of the future. Again, it did not spell the end to the traditions, cultural practices and social norms inherited from the past, but it did lead to a reshaping of that inheritance and the emergence of new forms of human relationships and acceptable usages as societies restructured themselves to meet the new needs. This was above all true of the changes which directly affected the working lives and family formations in modern society.

Men, women and children who had once worked the land, practised crafts or participated in cottage industries were displaced from their traditional means of livelihood and had to look for jobs in the factories and towns which had come to dominate the modern economy.

When we reflect on these past historical periods, it should come as no surprise that a similar kind of process is at work in our own time. Think of the gig economy. Still, the problems confronting the nations of the world at present are so vast and complex that they can seem at times as if they are impossible to solve, and so we unfortunately all too readily assume that to talk seriously of crisis is to indulge in pessimism, or that the only prophets that exist are prophets of doom. This assumption then becomes the most important dimension of crisis. It deludes us into thinking that the only choice of moods we have is that between optimism and pessimism, when in fact the real choice we must face is that between hope and fatalism.

A further difficulty that worsens the effect of this false dualism is that the serious challenges now facing us have taken place at a time in which a loosening of structures of 'collective identity' has occurred as well. Personal fulfilment in social, workforce, class, church and local group membership is not as prevalent today as it once was, and adult choices are now much more individualised. The Iphone and its apps like Facebook and Twitter are now key elements of people's "sense of belonging", instead of their local community networks. (if any still exist). As a result, one of the chief ways in which the current crisis affects us is that it complicates our lives. It introduces into our daily experience certain areas of conflict over which we have no direct control and concerning which all accepted solutions seem to fail. This does not mean, of course, that genuine solutions to social conflicts cannot actually be found, but it does mean that the eventual solutions will not be determined by simple 'local' actions but by the ways in which ordinary citizens respond to the much broader 'force of circumstances'.

This helps to explain why, at present, there is such intense, and even bitter, debate about whether the changes in morality, educational standards, sexual relationships, religious practice, lifestyle, cultural forms and modes of political protest have been for the better or the worse. The debates demonstrate that people are genuinely concerned about the changes affecting them and they are willing to become involved in public discussion about them even though nobody really seems to know exactly what the future holds in store for us. Finally, however, one thing is certain. As with any living species that is under threat of extinction because of internal disruption or external disharmony with its environment, the best chance of survival rests with those who are both capable of generating new energies within themselves and also are ready to adapt to the patterns of change transforming the world which sustains them.

Questions about sustainability and the need for a sense of balance therefore become central to survival, and are clearly tied to the growing concern over environmental issues – not only the type of changes that are often referred to under the heading of 'lifestyle', but also their flow-on effects already being felt within the industrial world as added cost. Pollution control, the side-effects of modern chemicals, the harmfulness of industrial processes to the workers involved on the production line in certain manufacturing industries, and the threatened extinction of a variety of animal species have all added to the safety control, public relations and legal costs of large corporations and the resultant dilemmas confronting political regimes throughout our world.

Because all this bears definite similarities with the previous cultural upheavals that our predecessors successfully survived, there may well be lessons to be drawn from the past about how the human race can now learn how to deal with the current problems it faces.

If we begin to develop for ourselves a genuine sense of learning from the past, this could help us cope with the mistakes we ourselves may have made in our own journey towards the future. We know our past, and that might, strangely enough, help us to move beyond the unknown and, for example, overcome our doubts about how to deal with our contemporary dilemmas such as Global Warming. Instead of deceiving ourselves all the time that "the past is over and done with", we can then avoid the temptation to think that true wisdom is to be found by holding on to the false dichotomy of "past versus present". What we need is a new Age of Enlightenment without the 18th century delusion that we had now become the 'Masters of the Universe' on the cusp of creating a Golden Age for ourselves and our children! Instead, this new age would be built on a return to a past, even pre-neolithic, view of ourselves as 'Custodians of the Planet'! We can learn a lot about this from the remaining native peoples throughout our contemporary world - in Australia and the Amazon for instance.

We may think that we know all about our past, and may believe that we have the means and will to share the world's resources, but the recent fires in the Amazon indicate that, even though we might think this, we are yet again following a mistaken path, and so there is still a lot of learning to do. A remarkable element in this is that the younger generation across the world appears to be much more aware of the mistakes we continue to make and even appears to be taking the lead in confronting the dilemmas of the future.

They are still young of course and cannot be excused from their own limitations or mistakes. But it remains true that young people seem to ask themselves instinctively: where do I belong in all this?

Political leaders and media commentators then express surprise and dismay when the young refuse to follow their lead and start to look elsewhere for their sense of identity. The young feel alienated, and even feel forced to "take sides" in the midst of all their uncertainty. Unfortunately, instead of recognising this as a genuine sign of a process of marginalization at work in the new generation, their elders read it instead as a sign of immaturity and then begin to blame them for being too simplistic in their approach to life.

For example, if we look back into the recent past, the 2005 racial riots in both France and Australia fed the popular diet of detached instant analysis, and few of the "adult" commentators were willing to face up to the factors of alienation that had finally found release in this 'sudden' upsurge of violence. If people - whatever their age - feel marginalized by their own society, it helps to be able to identify an enemy they feel that they can deal with or feel superior to, even if it means taking the law into their own hands. It is usually only a minority that pushes to this extreme, but the broader implications are there.

In more recent times the public protests about Global Warming by the younger generation, led by people like Greta Thunberg, seem to have been given the same treatment by many of those in power. As a result, and resisting the temptation to join in the chorus of "OK Boomer", one begins to wonder how the modern media would have covered past events that we now treasure as significant (e.g. the crusades, the medieval papacy, the burning of witches, the plagues, the journey of Columbus and the subsequent slaughter of the Incas of Peru, the Inquisition, the death of Thomas A'Beckett, the Reformation or the Napoleonic wars).

Would media coverage have been a caution to our predecessors as they gained access to video footage in documentaries to remind themselves of the fallibilities of the human condition? But, on reflection, in their own way the chroniclers of the past did do just that. What about Shakespeare, the later historians, artists, or even the composers of the operatic tragedies? Perhaps that is what 'the end of history' means in our own time; we are now free to repeat our mistakes.

So, to guard ourselves against this, we need to face up to the fact that we can still learn from our past. As was noted in *The Passage of Time*, instead of deceiving ourselves all the time that "the past is over and done with", we need to avoid the temptation to think that true wisdom is to be found by holding on to the false dichotomy of "past versus present".

Thinking of the past (history) should always inform attempts to predict the future (forecasting). This avoids the mistake of dealing with Time in a much more linear fashion – as a kind of trajectory, like a Spatial journey from point A to point B, which tends to reduce Progress to a process of 'leaving the past behind'.

We need to keep reminding ourselves that sometimes the clock needs to be reset, recharged, or even rewound!

3. THE CLIMATE HOAX

Leading politicians across the world are quick to dismiss current concerns as part of a 'climate hoax' being perpetrated nowadays across the world. Where is the evidence?

For example, why all the fuss about the recent 2019 Spring onwards bushfires across Australia, as if this is a completely new phenomenon, when all the historical records show devastating events have often occurred, not only during the many years of white settlement, but in pre-colonial times. In fact, in his recent book Dark Emu, the Aboriginal author Bruce Pascoe devotes a whole chapter to the first people's fire management traditions, dating back tens of thousands of years - removing vegetation, creating fire breaks, and the widespread use of one of their time-honoured traditional implements – the fire-stick! ERGO …. after all, if "we live in a sunburnt country", there is nothing new about bushfires in Australia and so it cannot have anything to do with post-industrial climate changes, despite what all the scientists and Fire and Rescue commissioners might say (eg, J. Gergis, *Sunburnt Country*) – how dare they.

Furthermore, while all this 'fire fuss' is happening in the Southern Hemisphere, the climate change propagandists are all carrying on about water, ice and floods in the Northern Hemisphere! Great Britain and Venice are paraded as key examples.

"Severe flood warnings and rail cancellations remain in areas of England flooded after a month's worth of rain fell in a single day. Derbyshire and South Yorkshire have been worst hit by the floods, which claimed the life of one woman swept away in a river near Matlock. Seven severe flood warnings - deemed a threat to life - remain on the River Don in South Yorkshire. Meanwhile, trains are not running in parts of the East Midlands".

And then there is Italy. "Venice's Tide Office said the peak tide of 1.5 meters hit just after 1:00pm but a weather front off the coast blocked southerly winds from the Adriatic Sea from pushing the tide to the predicted level of 1.6m. Still, it marked the third time since Tuesday night's 1.87m flood — the worst in 53 years — that water levels in Venice had topped 1.5m. Since records began in 1872, that level had never been reached even twice in one year, let alone three times in one week. While Venetians had a bit of relief, days of heavy rainfall and snowfall elsewhere in Italy swelled rivers to worrisome levels, triggered an avalanche in the Alps and saw dramatic rescues of people in countryside who couldn't flee rising waters". But, tidal swells, heavy rainfall and snowfall are not totally new, so why all the fuss? After all, back in 1966, long before all this 'climate change' talk, the tidal peak had been much higher, reaching a devastating 1.94m. And yet, over 50 years later the Venice mayor claims it is a direct result of climate change! Who is to be believed?

This brief section is titled The Climate Hoax, but the author is actually writing "tongue-in-cheek". It is an effort to get inside the mind of a climate-denier to see how easily we can stand the real evidence on its head and see it all as a huge hoax! Have we time yet to "beat around the bush", or "run for cover", or is the debate heating up now – not just the globe?

4. BECOMING CUSTODIANS

In *Facing the Future*, the proposed response to problems of global warming argued that it was now time for humans to cease viewing themselves as 'Masters of the Universe' and instead to prove themselves once again to be Custodians of the Planet. In our current globalised world this challenging transformation is likely to be more complicated than it looks. It is, however, the highly necessary starting point. To achieve it, we all need to start listening to what the marginalised people of the world have to tell us about the time-honoured values they believe have been sacrificed to the prevailing ideologies of domination. They might inspire others to find the courage to rescue those forgotten values from the Too Hard Basket.

Another way to express this is to ask ourselves what are the main human values that had to be sacrificed so that we could begin to adopt the 'Masters of the Universe' mindset.

One is obviously the attitude we adopt towards Time. This is a theme that formed a major part of the text of the *Facing the Future* section, which argued that thinking of the past (history) should always inform attempts to predict the future (forecasting). This avoids the mistake, identified in discussing *The Passage of Time*, of dealing with Time in a much more linear fashion – as a kind of trajectory, like a Spatial journey from point A to point B, which tends to reduce Progress to a process of 'leaving the past behind'. The Masters of the Universe try to create the future precisely by leaving the past behind.

It is informative to note that, in Australia, the disregarded First Peoples have a lot to teach the rest of us about this. For example, a recent book by one of them is titled *Sand Talk* and actually bears the sub-title *How Indigenous Thinking Can Save The World*! Maybe an extract could help us on our way. The author, Tyson Yunkaporta, discusses how we define Time, and refers to what he calls the indigenous First Law which attests that "creation is in a constant state of motion, and we must move with it as the custodial species or we will damage the system and doom ourselves" (*Sand Talk*, pp.45-6). Another indigenous book, *Dark Emu*, by Bruce Pascoe, provides evidence from the white colonists about the ways in which the First Peoples did act as the custodial species and yet were dismissed as mere 'savages' by the colonists who sedulously refused to learn anything from them about how to come to terms with living in such a strange continent which was so different from their own country of origin.

This in itself is a useful introduction to another key way in which we can rescue our lost values – owning up to our own mistakes and how our wrong choices have led us to develop closed minds about our need to change. Part of the mastery mindset many of us now share is likely to open up a can of worms for us here. Because that mindset is built around a series of 'either/ors' (dichotomies), each side is likely to be only too ready to supply us with a long list of 'mistakes' committed by their opposites (left versus right, male versus female, science versus art, West versus East, race against race, creationism versus evolution, regulation versus free markets, and us versus them).

Another way to explore this issue is to make use of the historical evidence provided by authors such as Hobsbawn in books such as *The Age of Extremes*, covering most of the last century, or White, whose study titled *A Short History of Progress* takes us way back into the past by making comparisons with our present. After examining the historical evidence he offers the following caution about current attitudes.

> The reform that is needed is not anti-capitalist, anti-American, or even deep environmentalist; it is simply the transition from short-term to long-term thinking. From recklessness and excess to moderation and the precautionary principle (p. 131).

This whole issue about 'mindsets' is becoming a crucial one and explains White's concern. The previous section about *The Climate Hoax* attempted to grapple with this and it proved to be a difficult issue to handle. One historian who has devoted considerable thought to this in a number of books is Yuval Harari, and in 2018 he even dared to offer us *21 Lessons for the 21st Century*. He highlighted how group think deceives us into thinking we are nearly always on top of the facts and also how 21st century technology becomes a substitute for thinking for ourselves and thus serves as a cocoon that shields us from unpleasant realities. He is even suggesting that we are not as smart as we think we are, as the following extracts show.

> A hunter-gatherer in the Stone Age knew how to make her own clothes, how to start a fire, how to hunt rabbits and how to escape lions. We think we know far more today, but as individuals, we actually know far less (p. 218).

People rarely appreciate their ignorance, because they lock themselves inside an echo chamber of like-minded friends and self-confirming news-feeds, where their beliefs are constantly reinforced and seldom challenged (p. 219).

Technology isn't bad. If you know what you want in life technology can help you get it. But if you don't know what you want in life, it will be all too easy for technology to shape your aims for you and take control of your life (p. 267).

Once again an Indigenous author provides the way out of this imprisoning mindset. In the second chapter of *Sand Talk* (on the theme of Forever Limited) Tyson Yunkaporta concludes

Any real move towards sustainability will require us to cease limiting our understanding with simplistic language around individual and group identities, villain and victim branding, so that we can see what our actual diversity looks like and what it can do for us (pp. 81-2).

Acceptance of diversity develops our ability to listen to, and learn from, different voices and escape the domination of being single-minded (or is that simple-minded?). The important insight to take note of here is that the acceptance of diversity is not simply a matter of replacing an 'either/or' with a 'both/and', but of accepting an element of uncertainty in how we face the future. Instead of seeing it as a question of dealing with a 'two-edged' quandary, it becomes a matter of learning to live with mystery as we try to shape a future for ourselves.

Ludwig Wittgenstein, expressed it this way: '… not how the world is, but that it is — that is the mystical'. In his earlier years he was very conscious of the limits affecting our search for meaning, and asserted that 'what lies on the other side of the limit will simply be nonsense'. Later he became his own best critic and suggested that this earlier view was based on a failure to look beyond the limits because of a mistaken desire for 'a preconceived idea of crystalline purity' (Dwyer, *Recovery*, p.63).

So, to make sense of climate change, we have to overcome our reluctance to live with uncertainty. As we go on our journey into the future, we may know what the important first steps are (and some of us may even already be taking them), but one thing holding others back is the 'masterful mindset' which insists on absolute certainty and a belief that mysteries should not worry us because now we know how to deal with them.

The mere mention of mystery might therefore be seen as a surrender. It does, nevertheless, provide us with the necessary key for opening up the Too Hard Basket. Learning to live with diversity (and the mysteries it confronts us with) may be personally challenging, and may even be seen as a 'cop-out', but the element of mystery that it involves means that a mere recourse to an algorithm or two will not be any salvation. Once we open it up, the Too Hard Basket is likely to contain the missing clues we need. To do that, we will need to learn to live with the uncertainty that mystery brings with it, and regard it as a challenge – not a disaster.

The big lesson from all this is that the first missing clue has now been revealed. It is that, for genuine custodians, dealing with uncertain outcomes has always been accepted as part of the job.

But it does not deter them from getting on with the job.

5. THE MINDSET ISSUE

Let us begin with a favourite quote of mine. It derives from around the time of the defeat of Napoleon at the battle of Waterloo. Back in 1814 a French intellectual, Henri de Saint-Simon, summed up the achievements of the Age of Enlightenment in the following way.

> Poetic imagination has put the Golden Age in the cradle of the human race, amid the ignorance and brutishness of primitive times; it is rather the Iron Age, which should be put there. The Golden Age of the human race is not behind us but before us; it lies in the perfection of the social order. Our ancestors never saw it; our children will one day arrive there; it is for us to clear the way (de Saint-Simon, 1814).

I have used this quote often because it sums up perfectly the mindset that underpins the whole philosophy of the Masters of the Universe: turning our backs on the past in order to 'clear the way' to achieve 'the perfection of the social order' and thus establish 'The Golden Age' for 'our children'! This single-minded trajectory is what leads the climate sceptics to reject the uncertainties that climate change introduces into our visions for the future.

As I noted in the course of *Becoming Custodians*, to make sense of climate change, we have to overcome our reluctance to live with uncertainty. I argued that, as we go on our journey into the future, we may know what the important first steps are (and some of us may even already be taking them), but one thing that is holding others back is the 'masterful mindset' which insists on absolute certainty and a belief that mysteries should not worry us because now we know how to deal with them.

The important insight to take note of here is that accepting an element of uncertainty in how we face the future helps us to live with mystery and regard it as a challenge – not a disaster.

It can also have the effect of making us more open to a diversity of views, and help us to understand that the climate sceptics are, in the words of Megan MacKenzie (Professor of Gender and War at the

University of Sydney), "not ill-informed or ignorant, they are just fragile and anxious". She even suggests that we can help them recover, by disentangling "the way that environmental degradation has been associated with masculinity". It all depends what suits them. Suits?? – that might even explain the Madrid COP25 outcome.

6. THE GALILEO EFFECT

In *The Human Clock* discussion of the issue of climate change, a suggestion was made that if we are to find effective solutions to our current problems we must be willing to 'turn back' the clock and learn from our past. If we were to do that, one of the significant issues we might dare to examine can be found about 400 years ago in the 1600s. It is the story of Galileo's fate at the hands of the Inquisition and the Roman Catholic hierarchy because of his criticism of a prevailing interpretation of biblical texts that placed the Earth at the centre of the universe. Instead his studies of astronomy (which even credited him with inventing up-dated telescopes) led him to claim that the Sun was the real centre, around which the Earth and other planets revolved. This view, which we take for granted today as a kind of universal truth based on reality, was in the 17th century tantamount to heresy.

Galileo lived from 1564 to 1642. He had even wanted to become a Catholic priest, but on his father's advice he began a medical course at the University of Pisa when he was 17. He soon gave this up because he developed a sudden interest in mathematics and science and in 1589, aged 25, he was awarded the Chair of Mathematics at the University of Pisa. He then moved to the University of Padua in northern Italy in 1592, settled there and earned his highly distinguished reputation for making many momentous scientific discoveries. By the age of 49 in 1613, his findings brought him into conflict with the prevailing wisdom, and all his written works were prohibited. In 1633 Galileo was interrogated by the Inquisition and threatened with torture. He was then sentenced to life imprisonment because he was "vehemently suspected of heresy." This was later lessened to house arrest, and 8 years later in 1642 he died, aged 77.

Despite his achievements, and his reputation today along with Kepler and Copernicus as a founder of modern science, his is quite a sad story that bears so much similarity to how, even today, a prevailing mindset can be so resistant to coming to terms with an unsettling reality. Or, to quote Al Gore, Galileo was guilty of promoting 'an inconvenient truth', but the lesson of history is that the truth finally won out. The Catholic Church did begin to stop banning some of his books, and eventually by 1835 his work was approved by the Church.

If we had lived in the 17th century we would have found it hard to believe that Galileo's findings would one day be regarded as undisputed fact – so there is a lesson in all this. Being up against the Masters of the Universe (whether they wear suits or clerical frocks) can seem a formidable task, but combining the lessons from the past with new insights about our contemporary experiences helps us to deal effectively with the uncertainties and mysteries that are so much part of life on this strange planet of ours. We need to care for it.

7. DREAMING LESSONS FOR CUSTODIANS

The Dreaming, an Aboriginal philosophy of life, has always linked Indigenous Australians with their land and what they accept as their Ancestral Spirits within it.

These Spirits once roamed the land, and the long-surviving stories about them acknowledge the life-forms and important geographic sites (such as Uluru) that they created, and which they continue to inhabit. These Spirit Ancestors include the Rainbow Serpent, the Mimi Spirits (fairy-like beings represented in cave paintings in Arnhem Land), and the Seven Sisters who represent the Pleiades star cluster.

What the Dreaming keeps alive (and is based on) is the inter-relation of all people and all things, and it thus helps to explain for all Aborigines the origin of the universe and the workings of nature and humanity. Because their Ancestors have remained in their sacred sites, the Dreaming is a never-ending reality, not a "Dreamtime" tied to a specific time, but an outlook linking people and the land, and the spiritual world of the past with the present and the future, It thus helps and allows Aboriginal people to understand their place in traditional society and nature, and thereby enables them to become genuine "custodians' of their world and thus fulfil their responsibilities to keep alive all they have inherited from the Ancestral Spirits.

Some of this is recorded in what they accept as Dreaming tracks, which join a number of sites tracing the path of one of the Ancestral Beings moving through the landscape, forming its features, creating its flora and fauna, and laying down the Laws which govern it all. An enormous responsibility – which surprisingly has survived, even though all this was taken over and suppressed by the colonial powers. The Spirit Ancestors perpetuate a long and distant past!

There is, therefore, a positive side to all this. For the Indigenous people, their custodianship has been proven right by the events of Climate Change which have left the colonial 'powers-that-be" at a loss. Not surprisingly, they have officially viewed the widespread droughts and bushfires from an economic perspective, and thus any attempts by ordinary citizens and, in particular, the original inhabitants to reveal the real links to problems of Global Warming have been dismissed as needless distractions. For the non-indigenous, there is therefore a big lesson to be learnt from the Dreaming: it is that all the elements of our planet are profoundly inter-connected.

This explains why, for the Indigenous, there is no such thing as <u>the</u> answer, because that would be a pretentious attempt to dismiss the complexities of our world by trying to be single-minded. So, if we make a genuine effort to see our world in the way that they always have, we might eventually realise, and even admit, that if we are ever to come to terms with Climate Change we must avoid the trap of finding <u>the</u> answer by excluding other explanations for these climactic events and thereby deceive ourselves into thinking that we can come up with a <u>single</u> solution. (cf Pascoe, p. 166)

The big lesson from all this is that, because of the tradition of The Dreaming, the Indigenous peoples continue their long-surviving responses to earthly events and that, after all, they do in fact know more about finding answers than the non-indigenous are willing to admit. It is time to accept their insights into problems like Climate Change that leave the rest of us perplexed.

BIBLIOGRAPHY

Berndt, Berndt, and Stanton (1982) *Aboriginal Australian Art: A Visual Perspective*. Sydney: Methuen.

Burling, R. (2005). *The Talking Ape*, Oxford University Press.

Dwyer, P.J (2010) *Life Journeys*, Sid Harta.

Dwyer, P.J, (2010), *Recovery: Towards a New Future*, Sid Harta.

Freire, P. (1970). *Pedagogy Of The Oppressed*, New York, Herder and Herder.

Gergis, J.(2018). *Sunburnt Country: The History and Future of Climate Change in Australia*, Melbourne University Press

Harari, Y. (2018) *21 Lessons for the21st Century*, Jonathon Cape

Hobsbawn, E. (1994), *The Age of Extremes*, NY Pantheon Books

Judt, T. (2010), *Ill Fares The Land*, Allen Lane.

Lewis-Williams, D. (2002), *The Mind In The Cave*, London,Thames & Hudson.

Mead, M. (1970), *Culture and Commitment*, New York, Doubleday.

Pascoe, B. (2014), *The Dark Emu*, Magabela Books.

Saint-Simon, H. de (1964), 'The Reorganisation of the European Community' in *Social Organisation, The Science of Man and Other Writings*, New York, Harper.

White, R (2004) A *Short History of Progress*, Text Publishing.

Yunkaporta, T. (2019), *Sand Talk*, Text Publishing Co.

Printed in the United States
By Bookmasters